CHRIST'S PURPLE HEART

KILLED IN ACTION BODY RESURRECTED

A

NARRATIVE

CHUCK HABERMEHL

RPF Publishing
ISBN 978-0-9725876-5-5

Proudly printed in the U.S.A.

I wish to thank the Lord Jesus Christ for giving me the greatest gifts of all: my family.

To my wonderful children, sweet grandson and the love of my life who designs all my books, my Angel. And Bean, my buddy.

I love you.

TABLE OF CONTENTS

PREFACE

Jesus is deserving of a special Purple Heart. My approach to this analysis examines the literature through the practical lens of a hand-to-hand, close quarters combative tactician. To ensure the recount of the final days of Christ's life was as accurate as possible, I found support in the study of The Bible, in numerous books on the life of Jesus and in period books on warfare in the region listed in the References.

Judea had been an occupied region for many years prior to Christ's birth. Roman leaders and Temple soldiers controlled everything that happened in and around Judea, including Jerusalem. It was Roman-conquered land and people, and the occupation life was marked by all things military: spies, mercenaries, weapons, soldiers, prisoners, cavalry, arrests, skirmishes and punishments, which often included the death penalty. The cruelest of all penalties was crucifixion.

Jesus had His marching orders in AD30, as you will read, with His operation planning and methods being similar to classic small unit tactics. There was the occupier and the occupied, the ally and the axis, good and evil. I have tried to document in detail, all of the injuries that Christ suffered through and endured on the day of His death. Four chapters are dedicated to the mechanisms of His dying: the interrogations, scourging, beatings and the crucifixion.

Many of the books and articles referenced apply most of their focus on the crucifixion. While this is where Christ expires, it is the massive trauma He receives beforehand from the brutal scourging and beatings that contributed heavily to His death. Of the mechanisms of His wounds, very little is ever mentioned about the specific beatings and bludgeoning to which He was subjected. It's almost as if they didn't occur. Yet it is documented that thrashings happened at several places: during His arrest, during at least four interrogations, while being force-marched to the soldiers' fort and anytime He was near the enemy.

In regard to the horrific physical tortures that Christ endured, I felt it was essential to elaborate on the wounds to His face, head and brain - descriptive injuries that are typically missing from most accounts of His death despite there being ample evidence from ancient scripture, recorded history and archeological documentation. Having experience with group dynamics made up of violent minds and taunting hostile crowds, it is no surprise what happened to Christ during a far more barbaric time where targeting vulnerable areas of the body, with and without weaponry, was commonplace. This inhumane treatment resulted in lacerations, punctures, bruises, swelling, contusions, hematomas and concussions. Except for a bone fracture, anything was fair play, for The Bible tells us, "not a bone was to be broken." And there wasn't.

By glossing over and watering down all of the physical suffering Jesus endured, the truthful violent reality of what happened to Him preceding His crucifixion is hidden. Perhaps this is one reason why a great deal of art and sculptures depicting Jesus on the cross do not fully render the graphic realism of what transpired. For everyone to fully appreciate the depth of the sacrifice made by Christ, it is important that His vile torture and death never be lessened in its unfathomable brutality and mutilation. The whole point is that He volunteered to accept all of the devasting trauma, to suffer as a mortal, in order to sacrifice His life for ours. The horrid abuse, with the sum of its repugnance, *is* the story. Without it, we lessen His sacrifice. It was necessary, and required, that Christ feel every bit of pain, misery and shame. Only this way would His sacrifice be sufficient enough to take away the sins of the world.

Jesus possessed all of the attributes of a warrior. He was a strong leader and great communicator. He was disciplined and was a disciplinarian. He was brave, exuded confidence, was unselfish and willing to sacrifice Himself for His unit, the mission and mankind.

On that fateful day, Christ was prepared to achieve His goal. In His combat preparedness, He had identified and knew the enemy. He trusted His allies and intel. The operations plan He developed covered small unit tactics, places for rendezvous, command and control, an area of operation, maneuvers and enemy activities.

I have witnessed all types of injuries to all parts of the anatomy during my time as a young Marine in Vietnam and during my lengthy career in law enforcement. While everything obviously pales in comparison to what Christ endured, I am familiar with what the human body, head and face endures from different types of blunt force trauma. There are resultant complications associated with multiple soft tissue injuries, including intense levels of mental stress, especially if one experiences a violent physical encounter, or knows He will be experiencing a violent physical death. Jesus Christ suffered such extreme heights of psychological stress that it resulted in Him sweating blood - a very rare condition that will be explained.

Periodically, you will notice the narrative shifts between the past and present tense to bring you closer to the critical moments as they unfold. I hope I was able to deliver a clear picture of what I believe is the detailed reality of Christ's sacrifice, and that you agree that there is no doubt Christ is deserving of a distinctive Purple Heart.

Chuck Habermehl
May 3, 2022

WARNING

THIS BOOK CONTAINS GRAPHIC CONTENT
THAT SOME MAY FIND DISTURBING.

INTRODUCTION

The Purple Heart Medal, as we know it today, is awarded to American military members who were wounded in combat and awarded to their families for those killed in battle. It did not exist in Jesus' time. His small detachment had no recognition or devices of any sort. Should Jesus be awarded a medal today, we can imagine what it might look like. While He's deserving of so many honors, including a Medal of Honor, it's because of His physical injuries that I selected a Purple Heart. But Christ's Purple Heart would have to be different, special.

As shown in Chapter 17, *Christ's Purple Heart*, I wanted to respect and preserve the shape and colors of purple and gold of the American Purple Heart awarded to hundreds of thousands of soldiers and marines, but preferred the main image on Christ's Purple Heart medal be distinguished not with an image of a specific leader, but with an image of the cross on which He died.

Above the heart is a shield, which symbolizes Jesus as the

protector. It is red for the sizeable amount of blood He shed and is flanked with thorns representing the crown and various devices used to inflict His injuries. The gold "V" that lies atop the shield stands for the valor Christ exhibited while being tortured as a prisoner of war. The red ribbon trimmed in purple is for additional blood He shed, more than any man could ever endure.

Judea was a Roman Province in 30 AD, with Jerusalem being the most significant city as it was the religious center of the region. This is where all of the Jewish leaders and The Temple Mount were located. Judea was formed and governed like all of Rome's conquered possessions. The threat of revolt and insurrection was always in the air. The occupied Jews hated the Romans who treated them as inferior. Likewise, the Roman occupiers hated the Jewish inhabitants. This was a time when religious zealots and other radicals were using guerrilla tactics against their Roman arch enemy. These disturbances were often widespread and threatened the fragile peace that existed. Robbers and thieves scattered throughout the province, attacked pilgrims and local travelers whenever the situation presented itself. Violence was common and widespread.

Occasionally, a wayward drifter, a charlatan, would enter Jerusalem and claim to be a prophet of God. This angered many of the citizens who were aligned with the elders. Eventually, the drifter would be exposed as a fraud, and after being chastised, sent on his way. Some were even given a physical lesson and suffered a beating or stoning for their forgery. These encounters intensified, as was evident following the baptism of a carpenter - a preacher named Jesus.

Jesus had been baptized on the banks of the Jordan River by John the Baptist, a legitimate prophet with a large following. Jesus' baptism initiated the escalation of His teachings. John the Baptist had recognized something special about Jesus and suspected, as did others, that Jesus might be the true prophet sent from God. Jesus began to develop a large following and this worried The Temple religious leaders who immediately put Jesus on their radar. Because John the Baptist was already a threat before Jesus was baptized, due to his popularity, John the Baptist was further harassed, spied on and labeled a threat by the religious leaders and marked for death.

Herod Antipas, one of the sons of King Herod the Great, had John the Baptist arrested. Although not a stranger to a prison cell, he would be killed by palace guards, who under the order of Antipas, beheaded him. Herod Antipas will also play a part in the outcome of Jesus' fate.

In *Christ's Purple Heart*, Jesus will be arrested, interrogated, beaten, scourged and crucified, and have many accomplices responsible for His death. This narrative will address the terrain; the theater of occupation; area of operations; operation order; the enemies and allies of Jesus; His arrest and the brutality of His punishments.

The chaptered details of Christ's ghastly wounds and the unbearable consequences of each torture during the final 18 hours of His life will elicit unspeakable pain and disfigure Him beyond human semblance.

✝

His appearance was so marred beyond human semblance, and His form beyond that of the sons of men.

Isaiah 52:14 (ESV)

BASELINE

To say it was quiet was an understatement, this time of night during the spring on the edge of the desert was calm and still. A barely audible shuffle blended in with the squad of twelve exhausted, weary, and now, hunted men. They slipped through the olive trees making their way back to Gethsemane, a rendezvous area they often used. Little did they know that a raid and prisoner snatch had been planned and a heavily armed force was just an hour away. With only two swords, the poorly equipped unit will not stand a chance.

In less than 24 hours the lives of the entire element will be transformed forever; eventually leading to the killing of all, but one. The first casualty will be the point man and team leader who has been classified as a high-value target. After He is captured and arrested as a prisoner of war, He will undergo brutal torture and suffer a vile death. This will initiate the birth of salvation.

PART I

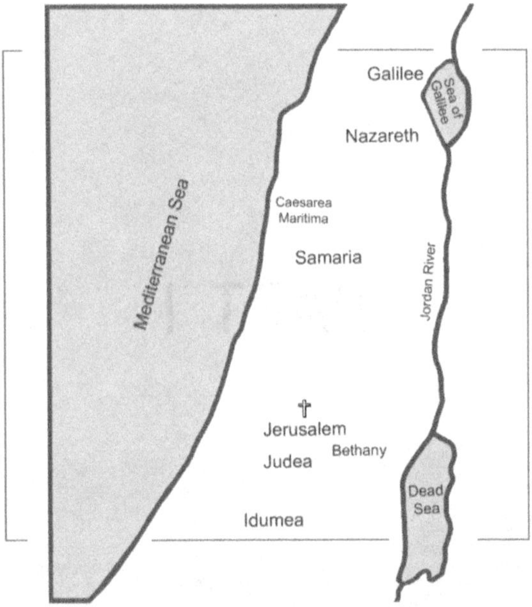
Galilee
Sea of Galilee
Nazareth
Mediterranean Sea
Caesarea Maritima
Samaria
Jordan River
Jerusalem
Bethany
Judea
Dead Sea
Idumea

Chapter I

THEATER OF OCCUPATION
(Judea)

Judea in 30 AD is an occupied territory. The entire geographical area is controlled by Rome. The Romans had taken the province and surrounding regions by force in 63 BC. By 6 AD the entire area was divided into three main districts: Samaria in the north, Judea central (including Jerusalem) and Idumea in the south. All the provinces and cities are complete with everything Roman, from various taxes, construction and trade, to government, troops (soldiers and cavalry), officials and laws.

Rome's way of appeasing the guarded population was by allowing them to maintain basic influence on local government doings and control over religious practices, laws and corresponding enforcement, minus the death penalty. With its northern and southern borders being Samaria and Idumea, respectively, Judea was bordered by the Mediterranean Sea to the west and, to a lesser extent, the Dead Sea on the eastern side. Terrain features vary widely in this mostly remote area, with deserts, hills, mountains, valleys, forests, coastal waterfronts, cities and villages. The elevation rises to a height of 3,346 feet above sea level on Mt. Hebron and falls to valley depths of 1,312 feet below sea level at the Dead Sea. Rainfall fluctuates throughout the province, depending on the terrain location characteristics and time of year. Although desert nights and higher elevations can be cold, the subtropic land is mostly hot.

Judea's hills are dotted with both small and large herds of grazing sheep and cattle. Herders and shepherds can be seen driving their livestock from one pasture to another. The mountains and hill areas abound with stone slopes creating natural steps. Caves are numerous. Travel by land is difficult due to extremely rough, rocky roadways. Only a few roads exist to connect the territory's larger cities and villages. Jesus is totally familiar with every part of this region. He has lived and worked in the area His entire life.

There is a large seaport on the Mediterranean named Caesarea Maritima. This busy port continually hosts small boats and large ships alike and serves as the gateway for the massive influx of trade goods that arrive daily. The seaport is always congested with pilgrims, but the numbers will swell beyond capacity during the week leading up to Passover.

The top official and overall authoritarian in Judea is the Roman Prefect Pontius Pilate. He has served at this post since 26 AD. He will remain in power until 36 AD. Pilate resides in the seaside resort Caesarea. His wife is with him at this mansion as is his complement of soldiers. Troops at this location consist of one cohort. This number amounts to 480 soldiers with a contingent of cavalry.

Rome has made a sizeable footprint in Judea. That is their objective and common practice - to have successful invasions and grow their footprint. Judea is the home of 7,000 auxiliary Roman soldiers that are strategically spread-out through the province as needed. Law enforcement in the area, except for Jerusalem, is the responsibility of the Prefect. Although Judea belongs to the province of Syria, each is governed separately. Judea is the theater of Roman occupation; however, Jesus will devise a mission that will take place in only one city. This chosen city is Jerusalem, the Jewish religious center of the world. Jerusalem will be His AO, area of operation.

✝

You will be my witnesses in Jerusalem and in all Judea and Samaria, and to the end of the earth.

Acts 1:8 (ESV)

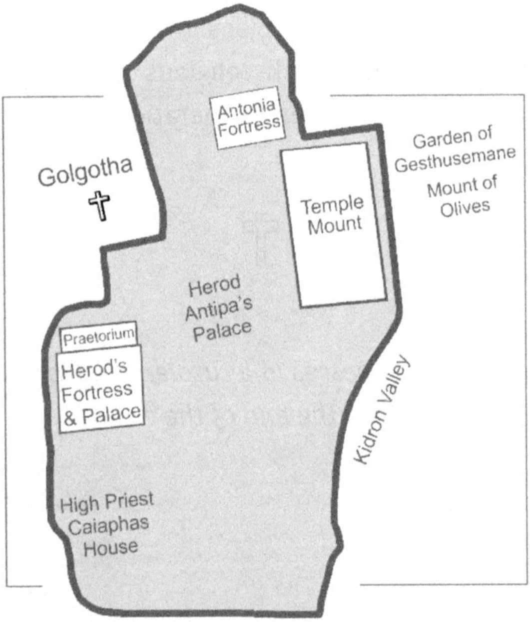
Antonia Fortress
Golgotha
Garden of Gesthusemane
Mount of Olives
Temple Mount
Herod Antipa's Palace
Praetorium
Herod's Fortress & Palace
Kidron Valley
High Priest Caiaphas House

Chapter II

AREA OF OPERATIONS
(Jerusalem)

The Holy City, the Jewish religious capital, a provincial city in Judea and a property of Rome; Jerusalem is Jesus' AO. This will be the area of operations for the mission that Jesus will carry out during Passover, 30 AD.

Jesus chooses Jerusalem for two reasons. First, it has a large population of 50,000, that will balloon to more than 250,000 during Passover festivities. This will allow Jesus to preach and teach to a multitude of worshippers, all who are hungry for enlightenment and need to be informed of the current scandalous government and their false religious declarations. Second, Jerusalem is the home of the only Jewish Temple in the world, and this means that religious leaders from all over will be in attendance, allowing Jesus to target vast numbers of authoritarians.

Jerusalem is situated 33 miles east of the Mediterranean Sea. There are three valleys that surround most of the city. During His final week, Jesus will cross the Kidron Valley, east of the Temple. The mostly dry wadi, skirts along the bottom of the

Mount of Olives. Both areas will be important for the operation. Like all of Judea, the geography of the city is abundant with stone and is relatively barren. A single spring runs water sparingly southeast of the city. The slight water supply is provided from cisterns, wells and aqueducts. The few roads that lead to Jerusalem are poor. The constant traffic from locals, and now thousands of visitors, keep it in a constant state of disrepair. Caravans of donkeys, livestock and wagons, with tens of thousands of individual walkers, continue to move along rocky potholed pathways. These challenges won't prevent the influx of excited travelers from reaching the grounds of the Temple Mount.

The layout of the city is divided between the upper city on the west and the lower city to the east. Each side has its own bazaar on the busy market streets. Many side streets and alleys splinter off from the two main roads.

Craftsmen and merchants conduct their businesses in animal, stall-like, spaces all along the streets. Sales of clothing, leather goods, jewelry and pottery are available. Food stuff is abundant, and meat is sold along butcher's street. Goods are made locally and imported as well. Tanneries, although needed, are required to be established outside the city walls because of the accompanying flies and foul odors. Even pottery makers are also not permitted within the city due to excessive smoke.

The Temple is the focal point for everyone arriving in Jerusalem. The majority have arrived via ship or smaller boats, while others come by land. All parts of the world are represented including Gaul, Rome, Greece, Cyprus, Asia-Minor, Syria, Egypt, Ethiopia and parts of Arabia.

The Caesarea Maritima seaport which sits on the coast of the Mediterranean is 65 miles northwest of Jerusalem and serves as the capital of Judea. The Roman Governor has his palace there. He travels to Jerusalem for important functions or large gatherings. A sizeable unit of soldiers with cavalry, provide escort, to and from.

As a child, Jesus visited the Temple grounds with His parents on holidays. He was keenly observant of all His surroundings. He may have even already been aware of His destiny.

The Temple Mount in Jerusalem is the central location for Jesus' mission. The current Temple is the second one built and had been enlarged and beautified by Herod the Great in 20BC. It is a huge complex, and includes courtyards, meeting spaces, various rooms, storage areas, lavish quarters and the Antonia Fortress. The Temple is where the Jewish religious leaders convene to conduct business. It is also the home of the High Priest Joseph Caiaphas and his Father-in-Law Annas.

The courtyards are bustling with activity. There are hundreds of customers and merchants conducting business, money changers are making transactions, visitors are sightseeing, locals are interacting with friends and relatives. Soldiers and Temple guards are patrolling the areas for signs of crime or insurrection. Thousands of people are crowded in and around the Temple grounds. With the continuous noise of bleating sheep, bellowing cattle and chickens clucking, combined with yelling salesmen and people talking over each other, the noise is deafening.

Jerusalem is a hub for olive oil production, and the Mount of Olives is a well-known location. Olive oil is the main export for Jerusalem, unlike Judea, whose main export is stones for construction.

Jesus will spend many hours of His last days at the Temple grounds. His many enemies are on a constant lookout for His presence. They also know the city, just as well.

†

Then one of them, named Caiaphas, who was high priest that year, spoke up, "You know nothing at all! 'You don't realize that it is better for you that one man die for the people than that the whole nation perish.'"

John 11:49-53 (NIV)

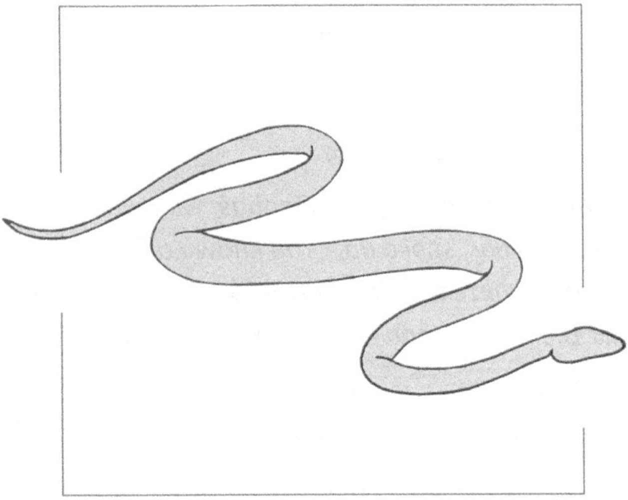

Chapter III

ENEMIES

Jesus detests the erroneous teachings, unlawful taxes, misuse of the Temple grounds and the scandalous relationships between the Jewish and Roman leaders. He is making Himself known, by those in charge. Those inside the Temple consider Him a troublemaker. His current sorties to Jerusalem from Galilee are becoming more frequent and His preaching more boisterous.

Jesus has a multitude of enemies that are in the midst of planning a strategy to have Him eliminated. A scheme that will finish Him off once and for all, like the way of John the Baptist. The Baptist had been beheaded by Palace guards.

Two men will ultimately carry the responsibility for the murder of Jesus, they are the High Priest Joseph Caiaphas and Judea's Prefect Pontius Pilate. There are, however, several others that aid and abet this crime: Herod Antipas, Annas, the Pharisees, the Sanhedrin, Scribes, Roman troops, Temple guards and the traitor Judas Iscariot.

Caiaphas:

Joseph Caiaphas, the High Priest, struts around Jerusalem like he owns the city; and in a way he does. He is the head of the Jewish religious leaders and as High Priest he enjoys all the perks that come with the job. His residence is the Palatial Mansion, a lavish section of the Temple surrounded by niceties and servants poised to respond to his every beck and call. Caiaphas is rich and arrogant. He is a hard man, one that deals harshly with local folk and pilgrims alike. His father-in-law is Annas, the former High Priest.

The organized corruption Caiaphas formulated transferred from one crook to the other. Bribery has been the best way to remain High Priest and Caiaphas, who is immoral and conniving, keeps the "hands greased" of the Roman leaders. He is possessed with greed and does away with anyone who interferes with his deals. An informal agreement exists between the two parties, with the leaders of the Romans and Jews allowed to benefit handsomely from unlawful taxes, fees, dues, financial penalties and graft. Caiaphas will remain in office from 18 to 36 AD.

It is Caiaphas that despises Jesus and wants Him killed. He believes his power is threatened by Jesus. He is becoming jealous of the claims that many are making in regard to Jesus' powers to heal the sick and perform other miracles. Caiaphas has labeled Jesus a fraud and considers Him a zealot and blasphemer. The charge of blasphemy alone is enough to receive the death penalty. The High Priest will insist incessantly that Christ be executed.

Pontius Pilate:

Pontius Pilate is the Prefect of Judea and governs all of the province. He has a large army at his disposal. Although he has jurisdiction throughout the region, he permits Caiaphas to control the city of Jerusalem. This reduces some of Pilate's responsibilities. Pontius Pilate is well known for his hatred and mistreatment of the Jewish population. He is described as corrupt, unyielding, ill-tempered, cruel and violent. His own soldiers consider him arrogant and tough. He is the only one in Judea with the authority and power to issue and carry out a death penalty sentence. Pilate benefits from his illicit relationship with Caiaphas and has tolerance for the priest as long as payments are adequate and frequent. Pilate will order that Christ be scourged and then crucified even though he admits he can find "no fault in this man." Pontius Pilate will serve as Governor of Judea and Samaria from 26 AD to 36 AD.

Annas:

Annas is the previous High Priest. He was in office from 6 AD to 15 AD. His daughter is married to Caiaphas, and it was Annas that arranged for Caiaphas to be appointed as High Priest. This set up allows the organized crime to continue for the former leader. Annas is sixty years old in 30 AD and is a driving force behind the arrest and conviction of Jesus. Annas is hot tempered, corrupt, a hoarder of money and is only dedicated to himself and his immediate family. He is very wealthy and wants to stay that way. He also lives in the Palatial Mansion. Annas is complicit with the Roman rulers and it is suspected that he carries more weight and is more powerful than his son-in-law. The former High Priest will demand that Jesus be brought before him first, affording him to be the first one to interrogate the prisoner.

Herod Antipas:

Herod Antipas is the Tetrarch of Galilee and Perea. He rules from 4 BC to 39 AD. Like all the other Jewish Kings and leaders, he is a puppet of the Romans. He has his own army of 4,000 men. Antipas is a son of Herod the Great. His father is known for having his soldiers search for baby Jesus in Bethlehem; he did not want the "chosen one" as competition. When the child Jesus couldn't be found, he ordered all children under the age of two to be slain. Herod Antipas has his father's cruelty and is infamous for his orders to have John the Baptist beheaded. He is considered violent, lazy and prefers to use his time to socialize and feast. He is ruthless in governing. When Pilate discovers that Jesus is from Galilee, he will send Him before Herod with hopes of avoiding passing judgement. He wants Herod to take over jurisdiction of Christ's arrest.

Herod has heard about Jesus and wants to meet Him. When Jesus, now the prisoner, is delivered, Herod is only interested in having Him perform magic or, better still, maybe a miracle. Antipas finds no fault in Jesus; however, Christ remains silent and refuses Herod's requests. This is an insult to the King. He orders his soldiers to dress Jesus in fancy garments to belittle Him. They mock Him, spit on Him and smack Him around. Failing to be complicit with Caiaphas or appease Pilate, Herod tires of the play and has Jesus returned to a surprised Pilate.

Sanhedrin:

The Sanhedrin is a group of ruling religious leaders in Jerusalem. This includes the Chief Priests, Scribes and powerful elders. Caiaphas, as the High Priest, presides over the Council and is forceful in insuring they do as he wishes. There are 71 members of the council, which includes Caiaphas. The Sanhedrin make the laws that govern all aspects of Jewish life. They act as the legislative branch of government that enacts laws; they act as the executive branch that enforces laws; and they also serve as the judicial branch that interprets the law. This is a convenient set up for all the members. There are no checks and balances of their conduct over the people. A large standing army of Temple guards receive their marching orders from Caiaphas and the Council.

They will all meet in the chamber of hewn stone for the Trial of Jesus. The Sanhedrin will vote the way of Annas and Caiaphas, finding Jesus guilty of blasphemy. They will demand He be crucified. After their sham verdict, they will all be outraged. Jesus is blindfolded and shackled, while each member will spit on Him and strike Him. Some punch Him and others slap Him, but they all take part in berating Him.

Pharisees:

Pharisees are the interpreters of Jewish law. They are a large body of both commoners and wealthy men who are very influential. They keep busy in all affairs of the Jewish government and find time to practice deceit and subversion. A contingent of Pharisees will confront Jesus at the Temple and attempt to entrap Him into making slanderous comments.

Some Pharisees will be part of the raid and capture team at the Garden of Gethsemane.

Sadducees:

Sadducees are an upper-class clan involved in the government. Most are wealthy. Although they are conservative, they will often disagree on religious issues with the Pharisees, except when it concerns Jesus. Many in this group are Chief Priests. The Sadducees also conspire to set Jesus up to commit blasphemy.

Scribes:

Scribes are involved in the interpretation of the Law of Moses. They are called lawyers and are considered legal experts. The Scribes are highly respected due to their level of knowledge. They too will join with all the other factions in attempting to implicate Jesus in having delivered false sermons. The Scribes and Pharisees are reproached for their corruption. They are supporters of the Roman occupation in order to keep graft going. They are also eager to eliminate Christ.

Judas Iscariot:

Judas Iscariot is one of the original twelve disciples. He is the only one who is not from Galilee but comes from Kerioth in another part of Judea. Judas will become the most well-known traitor in history. Currently, Judas has the role of treasurer of the group. He has helped himself to the coffer on several occasions. Most of the money he steals belongs to the disciples and Jesus;

the money was acquired through donations.

It will be during Passover week that Judas becomes disillusioned with the plans of Jesus. He has hoped that Jesus would gather enough support from the populace to over-throw the occupiers. This would create an independent kingdom, thereby giving Judas an important position. When he discovers the plan isn't about Jesus ruling a territory, he loses faith and joins the enemy side. His frustrations quickly turn to anger, and he approaches Caiaphas with a plan to betray his commander for thirty pieces of silver. This amounts to about four months' pay. Judas has agreed to lead the arrest team and raiding party to Jesus' location.

Jesus joins His disciples for a Last Supper in Jerusalem. Afterward, due to the late hour, He will stay near the city and will lead His men to the Garden of Gethsemane for the night. Judas, who did not make the trip, now knows where Christ can be found and the hostile plan is put in motion by, Caiaphas. A large contingent of Pharisees, Temple guards and Roman soldiers are dispatched to the Garden camp.

Roman Soldiers:

The Roman soldiers in Judea in 30 AD are all auxiliaries. This means they are recruited from the surrounding areas of Idumea, Samaria and Judea. There are some mercenaries that have been recruited to augment the manpower. The entire region consists of 7,000 soldiers. Judea itself has five infantry units and one cavalry unit. The combined force in Judea equals 3,000 men; however, this number will increase significantly during festivities, such as the upcoming Passover.

Jerusalem proper has only one cohort, the basic tactical complement for its population. It is stationed at the Antonia Fortress adjacent the Temple Mount. A Roman centurion commands the cohort. Pontius Pilate is the overall commander of troops throughout Judea. It will be a centurion, ordered by Pilate, to carry out Jesus' punishments. Knowing his men well, the centurion will designate the scourge team and the crucifixion detail.

Roman soldiers are known for their violence and aggressiveness. They hate the Jews they control and are always heavy handed during any encounters. The soldiers are extremely skilled in combat via real-world experience and no-nonsense training. They are hard men. Jesus will be on the receiving end of these killers throughout His ordeal as a POW.

✝

*Whoever wants to be my disciple must
deny themselves and take up their
cross daily and follow me.*

Mark 8:34 (NIV)

Chapter IV

ALLIES

Jesus is in His mid-thirties and in the prime of His life. He walks many miles a day and is a carpenter by trade. He is lean, strong and in perfect health. And He is aware that He will require dependable men to accomplish the goals of His mission.

Jesus has eleven dependable operators, His disciples. Usually, they can be counted on to do as He orders, but during the upcoming enemy raid they will escape and evade into the olive grove, abandoning Him to the enemy. Even His team leader, Simon Peter, His most trusted, will deny that he knows Jesus on three different occasions.

The number of Jesus' followers has grown over the past months and several of them have assisted Him in His activities in Jerusalem. However, due to their fear of the authorities, no rescue or help will materialize.

Many citizens of Judea and Jerusalem are wary of the High Priest and his counsel. They know that any contact with Jesus make them guilty by association. While they want to assist Him, they have to protect themselves, their families and businesses. They trust Jesus, but their fear of being persecuted is far greater.

Jesus will face the enemy alone.

✝

Then He took the twelve aside and said to them, "Behold, we are going up to Jerusalem, and all things which are written through the prophets about the Son and Man will be accomplished."

Luke 18:31 (NASB)

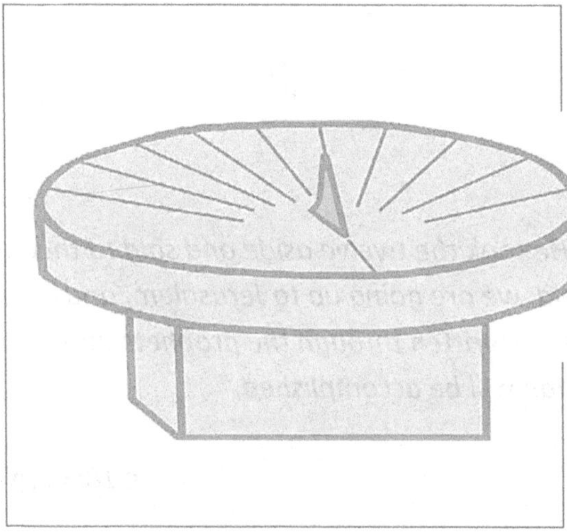

Chapter V

OPERATION ORDER

OPS Name: Operation Bedlam
Location: Jerusalem Temple grounds
D-day: April 4 - 5, Monday - Tuesday, 30 AD
H-hour: 0900

Unit Members: Jesus Christ - Element Commander; Simon Peter - Team Leader; James - Asst. Team Leader

Operators: Andrew, John, Philip, Bartholomew, Thomas, Matthew, James (Son of Alphaeus), Thaddaeus, Simon the Cananean

Situation: There are illegal and sacrilegious behaviors occurring on the Temple grounds, to wit, money changers conducting transactions illegally and over charging fees. Herders are driving their livestock through the Temple grounds creating chaos and filth. Sheep, cattle and fowl are being sold beyond fair prices and dirtying the floor. Various goods are being bartered illegally. The sanctity of the Temple is being disrupted and continues without interference from the corrupt Temple guards.

Intel: Illegal activities occur daily during the week of Passover. Most infractions are happening in the Court of the Gentiles and Basilica. Corrupt dealings with merchants and officials are blatant. Enemy scouts are lurking throughout the venue. Hundreds of people are roaming in and around the area. Scores are involved in questionable behavior. Many Temple guards are inside the grounds and many Roman soldiers outside the walls. They are patrolling in two-man teams looking for signs of trouble.

Comments: Team leadership and the operators want Jesus to issue a "no-go" due to heavy enforcement activity.

Mission Statement: To disrupt the money changers and business of other merchants by lecturing them about their violations, stopping livestock sales, preaching the Gospel and healing the sick and disabled.

Execution and Assignments: Jesus will release all livestock and fowl, scattering them away from the Temple area. He will scold the herders and explain their immoral conduct and related violations. He will turn over the money changer tables and put an end to their enterprise; interrupt their illegal business transactions by causing a disturbance; heal the sick and disabled; and preach to all. Team Leader and Operators will divide into three teams and spread out to different sections of the Temple grounds. They will keep a constant lookout for anyone who may try to prevent the mission and run interference against any soldiers or guards attempting to apprehend the team commander. Team Leaders and Operators will notify Jesus immediately of any suspected problems and perform area recon prior to initiation of operation.

Maneuver: The team will enter from the Kidron Valley, south of the Temple Mount. Simon Peter will choose a gate to enter that is absent of large guard presence, based on earlier reconnaissance. Teams will spread out into the crowd. At least three sorties from Bethany to Jerusalem will be conducted.

Logistics: Mary and Martha will assist as needed. Lazarus' home in Bethany (50-minute walk from Jerusalem) will serve as a fallback/safe house. The rally point and camp will be Gethsemane. There will be no field Command Post (CP), however, the mobile CP is wherever Jesus is located. Begin movement at first light and exit Temple no later than dusk, if possible.

✝

And Jesus went into the Temple of God, and cast out all them that sold and bought in the temple, and overthrew the tables of the moneychangers, and the seats of them that sold doves; and said unto them, It is written, My house shall be called the house of prayer; but ye have made it a den of thieves.

Matthew 21:12-13 (KJV)

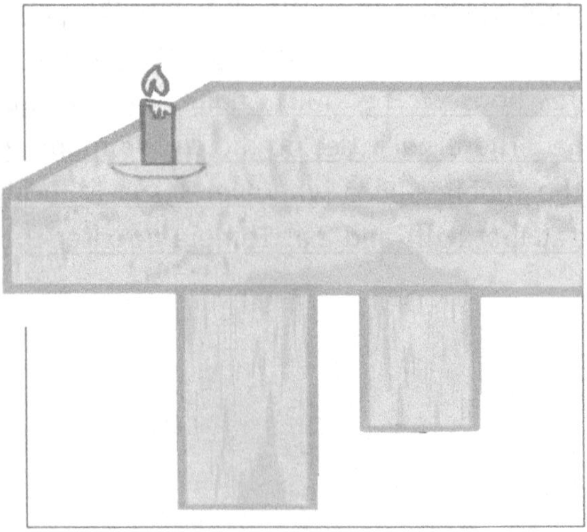

Chapter VI

AFTER ACTION REPORT
(Operation Bedlam)

Jesus and the team carried out their assigned missions pursuant to the operation order. Money changer tables were overturned. The racketeers were shocked. They were dismayed at what just happened and complained to the guards.

Numerous head of cattle and sheep were released and scattered to the countryside. Cages of fowl, mostly doves, were released. This created an uproar and many businesses ceased sales. Jesus scolded several merchants.

Several disabled and ill pilgrims were brought before Jesus, and He healed them and made them whole. This created a huge crowd that was loud and joyous over the miracles performed. Jesus began to preach loudly, proclaiming His disgust of the sinful ways being allowed at the Temple.

Lookouts alerted some guards and several Pharisees, who soon arrived. The Pharisees questioned Jesus and He cleverly rebuked them. The High Priest Caiaphas was alerted and apprised of what had transpired. He was livid.

Guards and pharisees report their version of Jesus' conduct to Caiaphas.

Caiaphas' mind is made up. He will waste no time in dealing with the troublemaker who is at risk of spoiling their wealth and control.

Then the chief priests, the scribes, and the elders of the people assembled at the palace of the high priest, who was called Caiaphas, and plotted to take Jesus by trickery and kill Him.

Matthew 26:3-4 (NKJV)

Chapter VII

CATALYSTS

The authorities are angry and want something done. Caiaphas has summoned the Sanhedrin and Jesus is targeted for immediate death.

Jesus is aware that His mission has been a success. He is also aware that He has a multitude of enemies that will now be after Him. In fact, at this very minute, they are in the middle of preparing and planning a strategy. A course of action is in the making to get rid of the zealot, once and for all, similarly, to the way of the "Baptist," but far worse.

Caiaphas and his Sanhedrin have been logging the discretions and preaching of Jesus, ever since he increased in popularity. The list is long. But there are three recent transgressions committed by the young rabble-rouser that have elevated His place on the religious leaders' *Most Wanted List*. It will be these three exploits that will serve as the catalysts for them to proceed with an immediate and aggressive hostile plan.

The first specific incitement that alarms the High Priest and the

Council is the news that Christ has supposedly raised a man named Lazarus from the dead. Even more shocking are the reports that the man had been dead for four days when this marvel occurred. Many witnesses have already confirmed the spectacle. Followers are beginning to call Jesus the Messiah. This news coupled with His growing number of believers has made the Jewish leaders become increasingly worried about His popularity and the possible repercussions it may have on the governance of the city.

Jesus is on the road traveling and approaching Jerusalem to attend Passover. He is flanked, on all sides, by enormous crowds. They have heard of His many miracles, including the recent and almost unbelievable tale of raising a dead man who had been expired for several days.

Crowds are jubilant and calling Jesus the Son of God and the True Messiah. He is even given a ride on the back of a young donkey and led to the Holy City. This is a sure sign that He is being identified as a true prophet. This is the second catalyst that reiterates to Caiaphas and the Sanhedrin that they must be expedient in their plan to eliminate Jesus. His behavior could start an uprising.

The final straw and third catalyst that moves the Jewish leaders to act, is Christ's recent performance in the Temple grounds. How dare He disrupt business, release animals to the wild and preach as a blasphemer. He created pandemonium by overturning tables and driving out customers and merchants.

Jesus also resisted being stumped by the legal and religious experts when questions were thrown at Him by the Pharisees and Sadducees. He has preached against Jewish Law and forced the High Priest's hand.

Extremely jealous, and worried about a revolt, the 71 members of the Sanhedrin will now develop tactics, establish a time frame and begin a course of action for Jesus must be murdered. How He must die has already been established.

✝

Then cometh Jesus with them unto a place called Gethsemane, and saith unto the disciples, Sit ye here, while I go and pray yonder.

Matthew 26:36 (KJV)

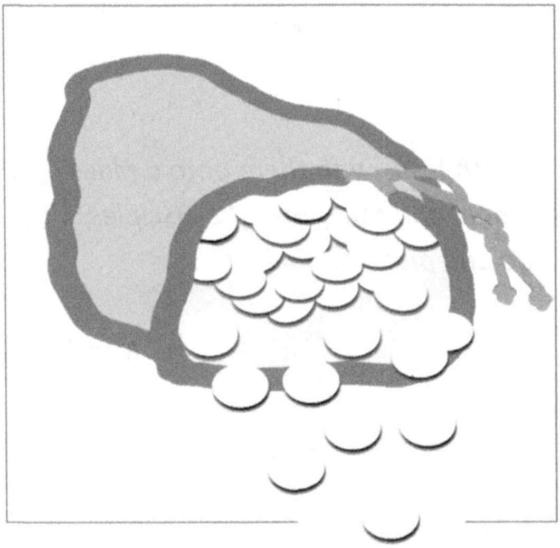

Chapter VIII

AMBUSH

Caiaphas summoned the Sanhedrin members to his house immediately after he had received the latest report of Jesus' conduct at the Temple grounds. Caiaphas is in a sour mood. The High Priest announces that the blasphemer will be captured and given a trial at once. He is not open to debating this topic and isn't asking the council's permission; he is essentially ordering them to agree. The decision is quick. The Sanhedrin is of the same mind, so when Caiaphas calls for the death penalty by crucifixion, everyone in attendance gives the green light.

Judas has made up his mind to betray Christ. His defection to the enemy has taken place. He has met with the High Priest and has made a deal that he will lead the arrest team to the bivouac site at the Mount of Olives. Judas will be paid the agreed upon sum of thirty pieces of silver upon delivery of Jesus.

The Temple Guard captain assembles his men promptly. Speed is of the essence. No one wants to be responsible for letting this fake Messiah get away. A platoon of Roman soldiers, Temple guards, several Pharisees, and even a couple of Caiaphas' private servants, set out with Judas directing the way. The raiders are equipped with torches, lamps, weapons and shackles with chains. One of the servants is Malchus, a man Caiaphas considers trustworthy and who will inform Him of any difficulties or blunders.

Jesus realizes that the enemy has ceased their surveillance for the day. Christ knows the time is right to exit the city. He silently leads His unit through some side streets of Jerusalem so as not to be seen. They have just finished the Last Supper. The night is cool and calm. A bright moon is shining as they cross the Kidron Valley. They are moving single file with a double arm's length between each man. The squad is hugging the shadows for camouflage. The shuffling of their sandals is the only audible trace of the team's presence.

Trudging up the hill, the Mount of Olives comes in view, and they blend in with the trees. They have arrived at the Garden of Gethsemane.

Upon arrival, the individual members spread out to previously assigned spaces. Before His leaders can settle, Jesus summons Peter, John and James. He instructs them to be on high alert and not to become distracted. Christ goes away from the rest, to pray to the Father.

Upon returning from His Prayer session, He discovers His three sentries, His most trusted, are fast asleep. They waken one by one. Jesus firmly dresses down the trio and accuses them of dereliction of duty. A warming fire is being stoked when suddenly, an ambush team, using complete speed, surprise and violence of action, dart into the camp area and demand everyone "freeze!" Startled and in fear for their lives, many of the disciples take flight and flee into the darkness. A few others stand and cower.

The ultimate traitor, Judas Iscariot, approaches Jesus and calls to Him, "Master," then gives Christ a kiss. This is the prearranged signal that the soldiers were waiting for; the wanted man has been positively identified.

Several assaulters grab Jesus roughly. Without hesitation, Peter swiftly draws his short sword from its sheath and slashes downward at the closest enemy to Christ.

The servant, Malchus, screams loudly and grabs at the side of his face where blood is already streaming. Peters defensive slash has sliced off the man's ear. Peter starts to continue with his protection of his commander, when Jesus commands him to stop and stand down.

Peter, confused, drops his sword. Jesus, under extreme stress, is now sweating blood. Stunned and confused at what they are all witnessing, they disregard the subject's condition. The raid is taking too long, and the ambush party is highly agitated. Jesus touches the wounded man's face and has reattached and healed his ear. There is no further resistance and, with the Disciples gone, the apprehension team starts escalating their use of force. Seizing Jesus tightly around the arms, His hands are shackled and chained, arms behind His back, elbows touching. A rope is pulled tight around His neck to prevent escape. The time is nearly 2100.

The enemy unit, with the prisoner in tow, retreats down the hill back towards the city. They are utilizing a secondary route to avoid running into any potential supporters. Without any witnesses present it is the perfect opportunity to teach the prisoner a few lessons. The platoon hurries the captive down the path, shoving, striking and dragging Him along. They are heading for the residence of the former High Priest Annas. He will be the first of four, to illegally question a man who is innocent of any crime.

Jesus is already in a weakened state from acute anxiety and constant worry. He is suffering from fretful anticipation of His upcoming ultimate sacrifice. He has just experienced hematidrosis while in the garden. Hematidrosis or blood sweat, is an extremely rare phenomena where blood drops can exit skin pores. It is brought on by the mechanisms of adrenaline, specifically the fight, flight or freeze syndrome.

The nervous system reacts to an imminent life or death crisis by secreting vast amounts of adrenaline. This is an automatic response to emotional trauma. The resultant adrenaline dump is unbelievably intense and causes a pounding heart, increased breathing, pale skin, profuse sweating, major panic and feelings of doom. If the hyper arousal and stress continues, then the small blood vessels near the sweat glands rupture and sweat and blood are excreted.

Christs' sweat glands have become dilated, and blood is being sent to His heart for survival. This is producing high blood pressure and rapid heart. Once the threat decreases and respiration returns to normal, blood rushes back to the skin and will exit the pores, causing Jesus to sweat blood.

Although this is a major mental and emotional strain, His blood loss is minor compared to the massive bloodletting yet to come during the next 18 hours.

Judas Iscariot has finally realized his sin. He attempts to give the silver back to Caiaphas to undo what he has done. Caiaphas refuses. Judas the traitor, who is the epitome of shame and deceit, commits suicide by hanging himself.

✝

Then they spat in His face and beat Him with their fists: and others slapped Him.

Matthew 26:67 (NASB)

PART II

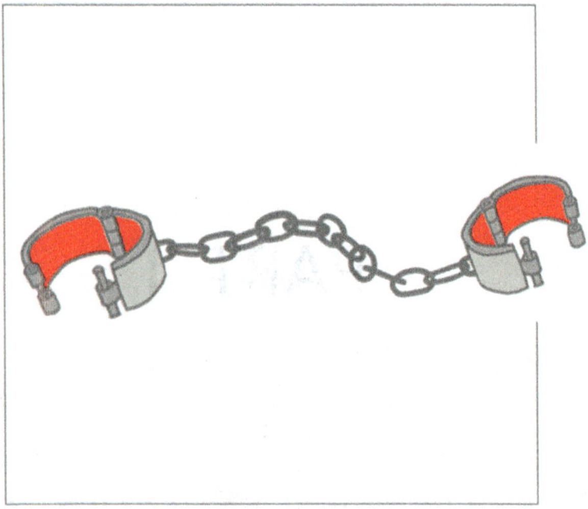

Chapter IX

INTERROGATIONS

Interrogation #1

The Temple guards have led Jesus to the Palatial Mansion, which is the home of Annas. Annas, though he is no longer the High Priest, is still involved in all matters affecting governance of Jerusalem.

Jesus is standing, still bound, arms behind His back before Annas. Numerous guards are adjacent the two. Annas poses questions which are not answered satisfactorily, if at all. Jesus knows He has been arrested illegally and that Jewish law is not being followed.

Not having any luck with the prisoner, Annas' scowl sends body language that the guards recognize all too well. Jesus is struck hard in the face and scorned by the guards. Treating *the shackled one* this way scores points with the boss. Frustrated, Annas sends Jesus to the main ringleader, Caiaphas.

Interrogation #2

After hearing that the POW is enroute, Caiaphas assembles the members of the Sanhedrin. This first trial is a sham. Trials are not supposed to be conducted in the middle of the night or without witnesses.

Christ, though fatigued, exhibits strength and defiance to the crowded room. The High Priest, acting cocky, barks questions and makes accusations in a rapid manner to the zealot. Jesus, still stoic, initially refuses to respond. He then incites the members, when He affirms to a question regarding His status as the Messiah. Without delay, the Sanhedrin and Caiaphas accuse and then adjudicate Jesus guilty of blasphemy. Caiaphas becomes so worked up that he tears his own robe in anger. The made-up charge of blasphemy is that Jesus said He is the Son of God. The verdict is rendered: Death.

Jesus is to be used as an example to other sinful preachers. He is ordered to undergo a punishment known for delivering the most pain and humiliation. Christ will be crucified.

The Sanhedrin and the High Priest are aware that this mock trial is not valid. They must hold a second one in the morning and they must also produce witnesses to testify against the accused. Even if they desire for Jesus to be crucified, it can't happen unless the Roman Prefect, Pontius Pilate, condemns and sentences the man. Jews cannot administer capital punishment. Only Pilate has this authority in Judea.

Despite these technicalities, Caiaphas and his large number of eager puppets are ecstatic that the fake Messiah will soon be eliminated. Their glee now turns violent as they begin to mock Jesus. This mocking quickly turns to striking.

Still shackled, chained and totally defenseless, Jesus becomes the punching bag of 71 members of the Sanhedrin and present Temple guards. This beating starts out with slaps across His face. Jesus is being covered with spit. Someone, probably Caiaphas, has a blindfold placed around Jesus' eyes. Now the real beat down begins, and he can't see the punches coming.

The soft tissue on Christ's face is swollen and red. Lacerations have formed along His lips. His eyebrows are swollen, as is His nose. He is bleeding. Heavily perspiring, He is becoming dehydrated. Rattled and injured by the scores of undefended punches, He can, at least, still hear and see. And for now, He is still recognizable.

The second farce of a trial commences at daybreak, pursuant to Jewish law. Several witnesses are called forward. These are transients and vagabonds that have been paid off by the elders with promised favors, threats or coins. Each utters lies. Nothing has changed, it is all an act, and once again Christ receives a death sentence.

Ragged, fatigued and in pain from His pummeling, Jesus is forcibly marched to Pontius Pilate for a hoped-for ratification of the death penalty.

Interrogation #3

Pontius Pilate resides in Caesarea. He lives in a lavish mansion near the sea. Like others, he is in Jerusalem to monitor the festivities during Passover. While here he will lodge in a section of King Herod's Royal Palace.

Jesus is taken to Pilate early in the morning immediately following the second trial. The workday starts early for the Prefect. This allows him to complete his administrative tasks before the sun rises high and turns the city into an oven.

Jewish law forbids entry into the Praetorium by the Religious Elders, so Pilate does a small favor and accommodates them by going outside to meet the gathering. Caiaphas explains the unanimous decision and position of the Sanhedrin. He then, almost begging, requests the Prefects backing. Pilate wants to interrogate this prisoner himself before agreeing to anything Caiaphas requests. The two men barely tolerate each other.

The subject, still handcuffed, is passed off to several Roman soldiers. His arms are still restrained behind His back, pulled tight, elbows touching, for more than nine hours now. His arms are numb. The soldiers shove Him to the Praetorium. Pontius Pilate has agreed to assess the man's guilt and then render a decision.

Jesus speaks cautiously to the Governor. Pilate is confused, he can't find any fault in the accused and furthermore, He appears harmless and non-threatening. This makes him suspicious of the motive of the ruling Jews. He is hesitant to support Caiaphas' request until he investigates further. Soon Pilate discovers that Jesus hails from Nazareth, in Galilee. He now sees that he may have an excuse to back out of having to ratify the arrestee's guilt.

Galilee is in the jurisdiction of Herod Antipas. Herod is the Tetrarch of Galilee and has full authority, like Pilate, in situations like this. It just so happens that Herod is in Jerusalem to catch the events. Pilate directs his soldiers and the guards to transfer the prisoner to the Hasmonean Palace at the Temple Mount and be handed off to Herod. Pilate sends a messenger to Herod to give him advanced notice.

Interrogation #4

Herod Antipas has heard of Jesus and is eager to accept Him in the Palace. Herod isn't concerned about any legal issues; he is hoping just to be amused by the so-called miracle worker.

Jesus is quiet. He remains silent and refuses to honor the King. Antipas won't get to see any magic tricks or observe Christ demonstrate one of His miracles. The ruler is insulted by this lowlife's refusal to entertain a King. Irritated, he plans to send the prisoner back to Pilate, but not before having a little fun.

Antipas has his bodyguards dress Jesus up in fine clothes to look more like a King. The guards smack Jesus around making Him stumble and fall while Herod laughs and mocks Him. Herod has eventually received his entertainment. He sends Jesus back to Pilate. The short trek includes more coarse handling. Christ has been without anything to eat or drink since yesterday evening, when He shared His last meal with the Disciples. There has been no end to His sweating. He is dehydrated and the random physical abuses are taking a toll.

During Christ's short trip to Antipas, a large boisterous crowd has gathered outside the Praetorium. Pilate is angry, but not surprised, to see the detainee reappear. The second evaluation by the Prefect doesn't go any better than the first. One thing is for sure, and that is Pontius Pilate still finds no guilt in the

POW. Pilate has made up his mind to release the harmless preacher. In fact, he will attempt to release the man on four different occasions during this day. But once again, Caiaphas has directed his underlings to incite the crowd; to only shout out for crucifixion.

Pilate has come up with an idea. He will release Jesus as part of the customary gesture to free a prisoner on a religious holiday. He is astonished and shocked when the frenzied crowd, worked up by the elders, asks for the release of the prisoner, Barabbas, a murderer, instead of Jesus, the giver of life. The anti-Christ mob is starting to turn hostile and begins chanting, "Crucify. Crucify. Crucify."

Pilate, feeling trapped, and concerned about a possible riot, gives in to the horde. He won't risk a civil disturbance. This would not sit well with Rome.

Pilate announces in a loud voice for all to hear, "Ibis ad Crucem." (You shall go to the cross.) The crowd cheers!

The Roman verdict is sedition, meaning Jesus has been inciting the population to rebel against the government.

The Prefect has one more card to play. He orders an extra severe scourging of the criminal. Maybe Caiaphas and the Sanhedrin will be content with a mangled and mutilated troublemaker. If so, the crucifixion can be stopped.

And they shall scourge Him and put Him to death: and the third day He shall rise again.

Luke 18:33 (KJV)

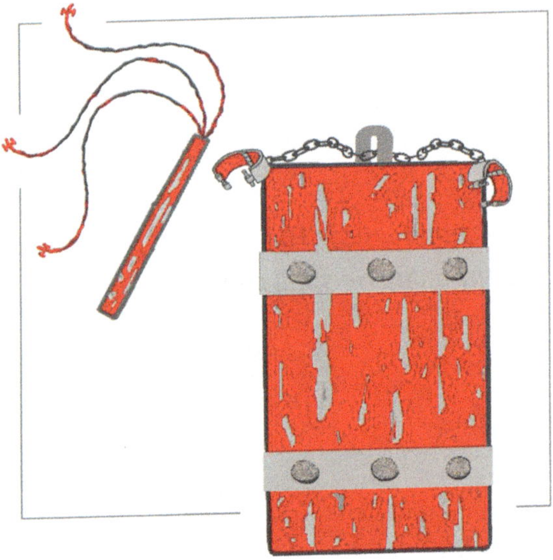

Chapter X

SCOURGING

Scourging or Roman flagellation was one of the most feared punishments in ancient times. It was often administered as a stand-alone torture and became an effective means of capital punishment. The whipping frequently caused death. In crucifixion, scourging was given in order to bring the convicted criminal close to death, which made time on the cross quicker.

The weapon specifically designed for scourging was the flagrum. There were several types of flagrums used in ancient times, some more wicked than others. Essentially, the basic whip had a short handle, wrapped in leather, to prevent the executioner's hand from slipping. Attached to the end of the handle were three to five, 18-to 24-inch strips of leather, called thongs. This was all that was needed to scourge a victim. Welts and lacerations would be delivered with every swing.

The configuration of the flagrum used for Christ's scourging was constructed to provide maximum damage to His body. This was achieved by attaching dumbbell shaped, hard metal pieces at the end of each thong. When a strike was delivered, chunks of meat will be removed. A ripping and tearing effect on the body. The

Romans used scourging on a regular basis. There was always someone who needed to be punished.

The Jews also used floggings to punish the accused. The difference, however, is that Jewish floggings had to be limited to 39 lashes; what they referenced as *40 minus one*. The Romans had no such restrictions on the number of lashes given.

The Romans were skilled at scourging. Scourging was more damaging than just whipping; so much more. The soldiers selected to the team were men who were powerfully built. They are recruited and assigned for their aggressiveness and are in excellent condition. These are important traits to prevent fatigue.

The soldiers that will flog Jesus today have vast experience and take pride in making their subject suffer and bleed. Collectively, they have flogged thousands of victims, or Roman criminals, in service to the Roman army.

Today they will try to outdo each other with each strike. Seeing who can cause the most pain. These men could care less about the accused. They are emotionally numb.

Security is tight in Herod's courtyard, where the punishment will take place. Even if the Disciples want to attempt a rescue to free their commander, it will be impossible.

Jesus is escorted to the yard. Finally, His shackles are removed. This is the first time in numerous hours that He can move His upper limbs. Unfortunately, His arms are frozen in position behind His back. The pain in His arms and shoulders is unbearable. Jesus has developed a condition known as hand palsy. This injury happens when blood flow is cut off or restricted to the hand by excessively tight shackles on the wrists, for a lengthy period. He has no sensation in His hands. Any reprieve is short; the floggers forcefully move His arms to His front. He winces in pain as He is stripped naked. By law, all clothing is removed. Complete nakedness is used to demean and humiliate. Padding is not approved, only bare skin. Once again, He is chained; this time to the scourge post.

The scourge post is a solid section of log placed deep in the ground to keep it sturdy. It can vary in height and is designed to stretch and expose already tensed muscles. This one is about four feet tall. Chains are hooked to the top of the post for the punishers to fasten the criminal in order to minimize His movement. This particular post forces Jesus to stand bent over and presents the floggers with a perfect target to maximize the strength and power of each strike, while reducing the fatigue on the team. Scourging has become an exact science for the Romans. It will be futile for Jesus to try and avoid any lashes.

The strikers take a firm, solid base. Their feet are spread at shoulder width apart. They place equal weight on both feet. This allows them to lean into the strike. They will use the rotation of their hips to produce more force from each swing.

An overseer is present to make sure no quarter is given to the condemned man, but he has nothing to worry about for two additional floggers are nearby in case a substitution is needed.

The torturer's stand on each side of Jesus and align themselves a few feet back, off to the side. This distance and angle will give full length of the thongs and metal pieces. Now in the set position, each swing can arch at max force targeting His back, shoulders, ribs, buttocks, thighs, limbs and the front of His torso. Some strokes will get His face.

Pilate has given his scourge team only one order: Don't let the prisoner die at the post. He must be alive enough to be crucified. Other than this stipulation, there are no limits.

The scourge begins.

A rhythm is soon established. Back and forth. Back and forth. Jesus screams at every hit. Every lash tear into His body. At first, little bits and pieces of flesh come off, then chunks of meat are removed. He cries out at each stroke. His body is being flayed open. Blood blisters, welts and broken blood vessels occur causing pooling bruises of blood under the skin. This is only the beginning. When the floggers land an especially fierce blow, where the strap takes a full bite out of the flesh, a grin appears on their face. When they spot an uninjured section of the body, they accurately target, with expert aim, and hit the area. To the soldiers, torture is a game. A game of horrific destruction. Whoever mutilates the most gets bragging rights.

Some in the crowd look away, others look on. Occupants of Roman cities are accustomed to violence; they have seen people scourged before but not with this much intensity or zeal. Christ's pleas and screams can barely be heard. He is hoarse. Christ is bleeding profusely now.

Their lashes land on top of previously torn skin and muscle. His back, which is absorbing many of the swings, is starting to resemble ground meat. Pain is unbearable and He tries to lose consciousness, but to no avail. Blood is flowing from His body and is creating red puddles around His bare feet that stream toward the punishers' sandals. A back-up scourger throws handfuls of dirt at his comrades' feet. They don't want to lose traction and diminish the onslaught. Jesus has received hundreds of extra brutal strikes because Pilate has ordered it so. He still thinks that a *good* scourge might sway Caiaphas and the others away from crucifying once they see their subject has turned to a bloody mush.

Jesus is suffering from massive blunt trauma. He is having excessive blood loss and has critical injuries. His breathing is shallow due to immense damage to His chest. Excruciating pain in His ribs comes with every blow, and now, every breath. His lungs are severely distressed making respiration almost impossible. His intercostal muscles between His ribs are shredded and hemorrhaging. The latissimus muscles and triceps are split open and bloodied.

Jesus is bleeding heavily in His chest cavity that creates undue pressure and causes one of His injured lungs to collapse. He is vomiting. Without any fluid intake, profuse sweating, for hours on end, and now significant blood loss, He is stricken with extreme dehydration. His mouth and throat are so dry He can no longer cry out. He now grunts when struck. His chest has been so beaten, creating a pneumothorax, and a ruptured spleen.

With each metal tip dug deep into what remains of His back,

Jesus' trunk is being broken down one swing at a time. His shape is changing. Christ is experiencing stages of hypovolemic shock; low blood volume brought on by excessive fluid loss. His breathing is rapid and labored. Blood in His chest is compressing His one inflated lung, inducing a post-traumatic pleural effusion. Circulatory complications are happening.

In order to survive, Jesus must breathe, but each breath delivers overwhelming pain. His injured lungs are delivering less oxygen to His body and struggling to remove carbon dioxide. He is coughing and vomiting blood. Christ is dizzy and confused. He is suffering from muscle cramps and spasms. Tissue and muscle hang from His body. He is alive, but barely. Not a moment too soon, the thrashing is halted by a Centurion.

Jesus is a bloody pulp. Large pieces of His torso are gone. He has numerous gashes on His buttocks and legs. He can't stand on His own. He is assisted to His feet and half dragged back to the Praetorium. He will wait here for the final journey.

Jesus' scourging blood bath, by and large, was an execution in and of itself. While waiting for the cross, He will be beaten senseless.

✝

And the soldiers twisted a crown of thorns and put it on His head, and they put on Him a purple robe. Then they said, "Hail, King of the Jews!" And they struck Him with their hands.

John 19:2-3 (NKJV)

Chapter XI

BEATINGS

Jesus is placed outside the Antonia Fortress to await His death march. The fortress is the headquarters for the Roman soldiers. Today there are more soldiers than usual because of the upcoming Jewish Passover bringing in so many pilgrims. With a constant rotation of men in and out, due to the many assignments and shift changes, the fortress area is packed by all ranks.

Hundreds of soldiers have gathered. They know what is about to happen and want to participate and cheer it on. Some were at the scourging, others are performing various duties, but most are milling around. Boredom is a substantial part of military life.

The Roman soldier has become increasingly more vicious as the years have passed. The army's response to years of conflict and wars filled with violent brutality is to escalate fighting with even more brutality and violence. Fight fire with more fire. The main reason Rome has been so successful in conquering other territories is because of the earned reputation of being ruthless in combat. Roman warfare before, during and after Christ's time, speaks of the legions conducting mass killings of men, women and children, ravaging cities of all their animals. Beheadings, crucifixions and bludgeoning were commonplace.

The soldiers themselves enlisted in an army feared for its savagery at a time when the known world is barbaric. The enemies of Rome that stood in the way of their world expansion were ripe with populations that possessed fierce, untamed warriors. As the Roman war machine became more experienced in the ways of combat tactics, some developed and some they copied from others, they have been able to accumulate a vast arsenal of fighting skills. They are adept at killing and killing quickly. But Rome's soldiers hold another distinction. They have devised tactics on how to maim and torture. They have become very talented in drawing out death and they seem to enjoy it.

There exist other reasons that produce such blood-thirsty veterans: soldiering is difficult and tough with many hardships. Army training is continuous and realistic, always striving for perfection at every skill set. Merciless battles develop merciless men. They are desensitized to feeling pain and to the sights of carnage. Life has no value. Since childhood, young boys grow up in a society that is rampant with killing and bloodshed. Crucifixions, scourges, impalements, gladiator games...it's all they know.

As occupiers, Roman soldiers always despise the inhabitants. In Judea they hate the Jews. Torture is an important part of warfare. It is often used to discourage uprisings and serve as warnings to bandits. When administered to enemy spies or combatants, torture becomes an expedient way of extracting valuable intelligence on an adversary's plans. Jesus will be the example today for the band of soldiers and their torturous rage. He has already sampled what callous, hateful men can do. The unexpected beating to follow will defy humanity.

More than 500 soldiers are congregated around the Praetorium, gawking at the condemned man. They are joking with each other about His appearance and laughing loudly at the claim of being a king. Many of the soldiers are off-duty and are quenched with lots of wine. Jesus is seated on a stool. His torso is curled over His knees. There is no escape from His pain.

He is the center of attention. Verbal taunts and insults transition to physical abuse. Only a handful of men are involved initially, but then they enlist others to join in. Jesus is being slapped hard on the sides of His face. These are strong men, and each open hand nearly knocks Him from His seat. The slaps are coming with more power; each one followed by a round of cheer and applause. Dozens of onlookers now join in. Not being content with their current strikes, because they aren't injurious enough, punching with a tightly closed fist takes over. Christ is being jabbed in the face. A straight-line punch, targeting His nose, mouth, chin and brow. Some drunks will hit His forehead. This would hurt the fist of most men, but these aren't most men. Additional swelling and new injuries surface on His face, which was last beaten during the interrogations, but left unpummeled during the scourging just this morning. The swarm is toying with Him now. That is about to change.

A human's pain threshold is an interacting mechanism of the brain and body that is largely influenced by adrenaline levels. The higher the adrenaline level, the higher the pain threshold. The degree of tolerance a person has to pain is also based on other factors: conditioning level, training, fatigue, hydration, desensitization, mindset, past injuries and fresh injuries.

Early on, Jesus' pain tolerance is high, due to a large quantity of adrenaline dumping into His bloodstream at the garden camp. However, as the day and assaults continue, His adrenaline will exit His body along with His blood and fluid volume. In regard to hand-to-hand fighting, a person with a high-pain threshold can absorb and endure a lot of pain before they succumb. In contrast, the person with a low pain tolerance will feel the pain associated with every strike, no matter how slight. Jesus has already withstood the raid and rough housing at Gethsemane. He has been struck in the face and head repeatedly by the Sanhedrin. Every interrogation and forced march to and from the different locations included physical abuse. The scourging He just received, has all but killed Him. He is suffering from extreme fatigue, life threatening dehydration, massive blood loss and unbelievable pain. His entire body is hurt. His muscles are cramping and spasming uncontrollably. He can no longer focus. Jesus' staying power is at zero.

Normally, when someone detects a punch coming towards their head, they will take evasive action. This consists of moving the head away from the strike or bobbing. This evasive action is paramount in minimizing the amount of energy that will be absorbed by the target. This energy reduction occurs because the power of the punch is diminished as the head moves away. The energy dissipates. However, when one doesn't or can't take evasive action, then the head absorbs 100% of the incoming blow. Jesus never had the chance to escape maximum injury to His head or face. Each strike Jesus received was essentially the same as an unexpected *sucker punch* because He was unable to physically see and to effectively defend Himself. He was shackled behind His back throughout His apprehension and the Sanhedrin had Him blindfolded as they punched and slapped His face. But Jesus was also blinded by the burning of salty sweat in His eyes and from the massive swelling around His brow and lids. His eyes were practically closed. He can't see the hits coming so He doesn't know when He will be struck. No defensive action. This is when full force is delivered. His face and head absorb all the energy, causing maximum damage to the part of the anatomy struck. And His body part receiving the blow is not the only part susceptible to injury; surrounding structures are impacted as well. For Christ, this will include His brain, neck muscles and vertebrae, eyes and ears. The power and angle of attack, combined with the type of blow and follow through, will increase the damage to these parts.

When the head or face is struck with blunt-force trauma, the brain moves inside the skull. If the force of the hit is powerful enough, which in Jesus' case they are, the brain will bounce against the skull causing a traumatic brain injury. As Jesus is punched in His head and face, with a straight-line strike, His head snaps backward. This linear acceleration causes His brain to slam forward and strike the inside of the front of His skull; this is called a coup. When the head quickly returns to its preferred upright position, the brain then jolts rearward contacting the back of His skull; this is known as a contrecoup. The coup-contrecoup movements can cause hematomas, contusions, shearing, concussions and a traumatic brain injury. The punch fest is turning into a feeding frenzy, everyone wants to get in their licks; each trying to out punch the other. Christ has become a human punching bag. Even soldiers who had not planned on participating in the beating, join in so not to be considered weak. And as long as they follow the news they've heard - that the fake king must be killed by crucifixion only – any other abuses are fair game. The barbaric heathens have no empathy for Jesus. They enjoy the cruel assault against Him. So, the pummeling continues.

A few pagans are striking with a hooked punch, a circular acceleration strike that produces more force than the straight-line acceleration of the jab or cross punch. Roman soldiers are skilled in barehand fighting and know how to achieve maximum destruction by sometimes punching *through the target* - a destructive method where the punch reaches full extension after the target is hit. These men are godless, profane brutes; they are more like animals than humans.

Scores and scores of enlisted men are taking part. Jesus' face is grotesquely swollen now. Blood covers His face and beard. His eyebrows and eye lids have been lacerated. Blood is flowing into His eyes. Punches are landing in the same places again and again. Gashes are growing larger. Both of His eyes are swollen shut. He has most likely suffered injuries to His iris, cornea and retinas. His brow region has been struck repeatedly, creating cuts that are bleeding heavily. Three fifths of the nasal septum is not protected by bone, making the small vessels in His nose susceptible to injury. They have ruptured and blood is pouring. Much of the blood flow is streaming inside His throat making His already-compromised breathing worse. The lateral cartilage of

Jesus' nose is being reshaped. Tissue along the nasal cavity has separated and cuts appear on both sides of His nostrils. His mouth is full of blood and mucus. The targeting of His lips and jaw is resulting in soft tissue damage inside His mouth. Skin is being mashed and sliced against His teeth. He has multiple mouth injuries and a dislocation of His jaw.

Christ is being knocked off His stool repeatedly and forcibly thrown back on. Those who have injured their hands refuse to quit, and simply strike using different surfaces. Jesus' face is taking a tremendous beating. With these different methods of aggression, a rotational acceleration brain injury cannot be ruled out. Brain injury from rotational acceleration is more serious and harmful than linear acceleration. When blunt force hits the side of the head, the brain rotates within the skull. Because the brain is fixed in position to some degree, thread-like nerve axons can be torn, tissue sheared and blood vessels ruptured. Due to the magnitude of pounding that Christ has been exposed to, concussion and traumatic brain injury would, most certainly, have occurred, again. This would mean Jesus sustained a second impact injury to His brain – when an additional concussion happens before the first one has healed. Jesus is starting to feel His life fade away. He has undergone nonstop bludgeoning for almost an hour. He is no longer recognizable.

Not yet satisfied, a couple of goons dress Jesus up as a king. They make a crown from twisting and plaiting several thorny branches common to the Judean region and give Him a thick bamboo-like reed that is strong and flexible to be His scepter. They cloak Him with a purple robe. Christ has undergone such enormous physical abuse, yet the infliction is still not over.

More than one hundred soldiers are actively taking part in this horrific *event*. Five times that number are spectators. Jesus is wearing a robe, barely holding the scepter and has a crown of thorns shoved downward onto His head. His sweat-tangled hair clumps with fresh streaming blood from the piercing. The contingent of beasts go on with their entertainment only now they grab His scepter reed and swing it downward onto His head. Any contact with the crown drives the two-inch-long thorns deeper into His head and scalp. The thorns pierce his trigeminal nerve in his forehead and the occipital nerve branch in the back of His head. Significant bleeding occurs from these deep painful punctures that are almost indescribable.

Christ's matted hair, face and beard are caked with blood, spit and sweat. The reed strikes to His head have caused stabbing pains down His face, to His lips and nose, as well as additional trauma to His brain. Jesus has taken so much torture to His face, it is disfigured. The gruesome pounding finally comes to an end when the Centurion in charge of the crucifixion detail calls out for a halt.

Christ appears unconscious and is not moving. The onslaught wounded every part of His now fully bloodied face and head. No part has escaped damage. The first sign of life is when He vomits more blood. Jesus is dizzy, He is having trouble hearing and His eyes are still swollen closed. The wounds to His face and head are as atrocious as the scourge wounds that cover His torso. The only reason Christ has not already perished is because His sacrifice has not concluded.

Hypovolemic shock progresses. Jesus' breathing is shallow and rapid. Completely dehydrated, His fluid loss is at a fatal level.

He has lost huge volumes of blood and is close to organ failure. His body trembles, uncontrollably. Severe muscle cramps and spasms are attacking His limbs, back and stomach. What little skin can be seen through His blood, is ashen. Christ's overall appearance is repulsive and ghastly.

The centurion in charge of the crucifixion detail orders his complement of twelve soldiers to prepare Jesus and the other two imprisoned thieves for the trudge to Golgotha. They are all slated to be killed today, before dark, before the religious festivities begin.

Before the procession can begin, Pontius Pilate has summoned the condemned preacher to be returned. Pilate instructs two palace guards to exhibit Christ's ravaged and blood-soaked body to the masses. "Behold the man," Pilate announces to all present. The Prefect is hoping for the display of the horribly battered prisoner to be considered punishment enough and alter the crowd's decision. Just as Pilate initiates the question, He is interrupted and shouted down with "Crucify. Crucify. Crucify." This permanently puts an end to any other alternative. Christ, a sinless man, will be crucified.

✝

And He, bearing His cross, went out to a place called the Place of a Skull, which is called in Hebrew, Golgotha, where they crucified Him, and two others with Him, one on either side, and Jesus

John 19:17-18 (NKJV)

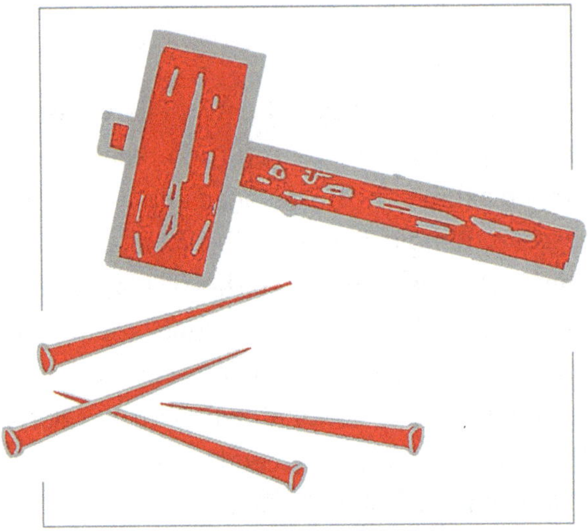

Chapter XII

PLACE OF A SKULL
(Golgotha)

The Place of a Skull is the interpreted name for Golgotha. There are some that believe this place received the name due to the shape of protruding rocks that resemble the shape of a skull. Others think that this death spot got its title because of all the crucifixions that are carried out here. Still others surmise that its name comes from the barren landscape and poor terrain. Regardless of its name, everyone in Judea knows that this busy patch of ground is only used to kill.

Golgotha is adjacent to Jerusalem, but sits outside the city walls separated by the Garden Gate. *The Skull* proper is rocky, dry wasteland.

Crucifixions draw sizeable crowds. The people make every effort to not miss such entertainment of a criminal's suffering. But this is an even larger crowd for there is heightened intrigue surrounding this particular punishment because one of the adjudicated is the self-proclaimed Messiah.

Pilate has instructed his commanders to beef up security and place reinforcements on standby. This will prevent any disruptions. The Place of the Skull will host its most famous mortal this day, April 8, 30 AD.

And it was the third hour, and they crucified Him.

Mark 15:25-39 (KJV)

PART III

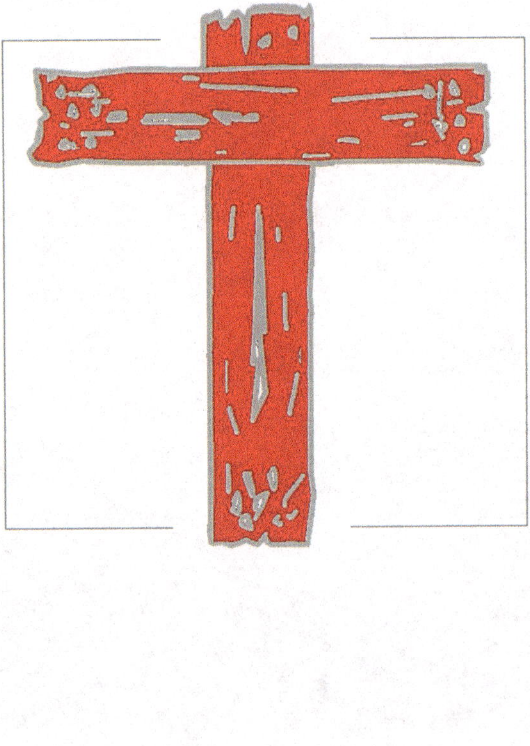

Chapter XIII

CRUCIFIXION

"Cruelest and most terrible death penalty"
Cicero, Roman Orator / Writer
106 BC - 43 BC

The Exactor Mortis is the Centurion who will command the crucifixion detail. He also orchestrates other death penalty verdicts. He is extremely efficient. This is a permanent assignment designated to a soldier who has proven his skill hundreds of times in the tasks that are necessary.

At first, the procession is arranged and placed in formation. The center section consists of the three subjects to be executed. They are lined up in single file. Jesus is in the middle; this will also be His placement at Golgotha. Each victim has four soldiers assigned to them who will do the dirty work. The four-man detail is strategically placed - one in front of the prisoner, one at the rear and a man on each side.

A small unit of Cavalry, mounted on two-thousand-pound war horses, lead the front and bring up the rear of the formation. They will keep the route clear and perform riot control duties if needed. A platoon of soldiers surrounds the center section to prevent any interference or attempted rescue. The Exactor

Mortis rides his stallion at the front of the center section and leads the way. Before moving forward, the horizontal section of the cross, the patibulum, must be hoisted on the criminals' shoulders. The Law demands the condemned carry their own cross. The patibulum weighs 60 pounds. Jesus is in such wretched condition He can't even stand unassisted, let alone, carry His cross piece.

The Exactor Mortis is considering altering the route to a more direct path, for fear that Jesus may die if they go the usual, longer way. However, the Romans preferred the more common and frequently used roads for the accused to be viewed by the most people which would serve as a deterrent and example to others. He will go the usual way.

The distance to Golgotha is just over a half mile of mostly narrow roads. An easy walk for the soldiers and unscourged, unbeaten bandits; an impossible task for a decimated, half-dead Jesus.

The sun is high overhead and shining bright. It is hot. Christ is still wearing the robe and crown of thorns. He is in unbelievable pain, He is cramping. He is dizzy, confused and feeling light-headed. The enormous fluid loss is starting to impact normal organ function. Christ can barely see or think. He is in a daze. Jesus is struggling to shoulder the cross beam. He has already staggered and fallen several times striking His face and head on the road; the crown's thorns are driven deeper. The formation has been halted repeatedly. The Centurion is furious. He will not allow this pseudo king to delay his duty. An onlooker, Simon of Cyrene, is quickly volunteered by the Centurion to carry the patibulum.

Three thick vertical posts, called stipes crucis, are already in the ground at the place of the skull. The stipes see a lot of use. The timbers are cracked and well-worn. Jesus is stripped of the robe. This yank opens clots causing dried wounds to bleed again. Two soldiers throw Him down on His back placing Him against the rough horizontal cross section. A soldier pushes down on the crown to ensure it won't fall off.

Crucifixion was widely used as capital punishment in the ancient world. Sources report that crucifixion was used long ago, by barbarians. The actual birth of this cruel punishment depends on the reference. Latin texts mention its use as far back as 218 BC; however, it has also been catalogued to 260 BC. Herodotus tells of a Persian General crucified in 479 BC. Regardless of the exact dating, we do know that it has been around a very long time and was a popular form of punishment. Crucifixion included scourging, beating and other tortures as part of the complete punishment. It is believed that the Carthaginians first passed the technique to the Romans. Like all things the Romans borrowed from other countries, they soon perfected its use. Rome used crucifixions as a political, civilian and military penalty. However, it was slaves, the poor, criminals and non-Roman citizens that were the usual victims. Roman citizens were supposedly exempt from this extreme wrath; some would discover differently.

Crucifixion is heralded for its efficacy as a valuable deterrent to committing a crime or a population revolt. It is considered the most horrific of all sentences. It outdoes beheading, burning at the stake, being eaten by wild animals and even the gruesome impalement. Impalement consisted of driving a long-sharpened stake in the criminal's rectum, upwards through the torso and out the mouth. This inhumane demise was considered very effective but had one major flaw; it killed too fast. Crucifixion was considered the ultimate penalty. Not only was it horrifying to witness the brutality that delivered the unimaginable pain, but it took several hours or even days for the subject to perish. This magnified and extended the torture.

In addition to being the cruelest and most severe method of death, it also delivered humiliation, which was extremely important with Roman executions. The unlucky victims not only died slowly, undergoing immense suffering, their naked bodies were on display, exposed to the elements, animals, family and foe. A complete and lengthy insulting conclusion to their life.

The Romans tortured and crucified prisoners completely naked, to increase the indignity and shame. There may have been some circumstance where a condemned Jewish prisoner may have been allowed a loin cloth, but there is no reason to believe this would have been granted to Jesus; Caiaphas possessed too much hatred for Him and wanted ultimate pain and the utmost humiliation.

The Gospels didn't have to explain much about crucifixion or any of its related horrors because citizens of the era knew all about them from their frequent first-hand viewing. Rome's leaders seemed to enjoy handing out the punishment. Consider the 6,000 slaves crucified during one instance. They were captured after the failure of Spartacus' slave revolt. It was said that crosses lined the road for 120 miles in Rome from the Appian Way to Capua. The soldiers had difficulty in finding enough wood to make crosses.

Crucifixion remained popular long after Christ's murder. In 66 AD, in Jerusalem, 3,600 Jewish men, women and children were scourged and crucified on the orders of the Roman Governor Gessius Florus. This was to serve as an example to prevent future unrest. The Jewish inhabitants had become angry at the governor's stealing of gold from the Temple. Roman Emperor Constantine the Great, the first Christian Emperor, abolished the punishment of crucifixion in the year 337.

Caiaphas has a lust for revenge against Jesus. He wants the manner of death to serve as a deterrent to any others, groups or individuals, who might consider challenging his power. He wants Jesus to suffer.

The four-man execution team assigned to Jesus went to work right away. After He was thrown on His back, one soldier holds His right arm down, then the left, stretched out along the patibulum, keeping the arm still, another soldier straddles Christ's torn chest, another on His battered legs. The main executioner is stocky and muscular. He is carrying a heavy mallet in his right calloused hand and four, six-inch long, rusty iron spikes in the other.

The executioner places the thin blunt end of the spike near the base of Christ's palm; a spot called the Thenar Furrow. The thenar furrow is the fleshy part of the palm near the thumb. Using an overhead swing he smashes downward driving the dull spike downward at an angle towards the wrist. After two more hits the spike exits a space between the ulna bone and radius bone and enters the crossbeam. Years of killing experience has taught the Romans that spikes nailed into the middle of the palm will not hold the weight of a man and will instead tear away. The other hand and wrist are nailed in identical fashion. The large spikes slice through the median nerves of Christ's hand. Somehow, nearly lifeless, He screams in pain. This damaged nerve causes causalgia, an intolerable, continuous, burning pain - a pain so severe a state of shock ensues. Reports from World War I stated that even morphine was ineffective for soldiers afflicted with this type of wound. Making some take their own lives. Jesus will receive four of these unbearable punctures.

With Jesus nailed solidly on the patibulum the crew positions themselves to lift Him up and on to the stipes. Two squad members, one at each end, grab the horizontal post; the one in front of the body grabs Christ around the waist and the fourth man is ready to lift His feet. The Exactor Mortis shouts for the convicted to be lifted. In unison, all three victims are picked up and the patibulum's are lowered onto the stipes. The beams firmly lock into the pre-notched slots.

Jesus' every move and breath is wrought with excruciating pain. The squad bends Christ's unbendable legs. His feet are placed side by side with His soles flat on the stipes. The soldiers know this man won't offer any resistance; His condition is already dire. The second set of spikes are driven through the top, center of each foot near the instep. The spikes crash between the second and third metatarsal bones settling solidly in the beam. The planter nerves are cut sending shocking jolts of misery throughout Jesus' body. He thrashes and tries to scream. The enormous pain He has experienced during the past 15 hours has just intensified beyond belief.

Christ is now mounted on a cross on Golgotha.

There are several types of crosses in use in this era, with two main shapes. The one most often seen in art and literature is the Latin cross, a lower-case letter *t*. The other cross, the one Jesus is crucified on, is called the Tau cross, which is in the shape of a capital letter *T*.

It's time for the cross to complete the job that the scourge and beatings started. Christ hangs, lethargic, nearly comatose. He faintly hears the jeering coming from the mockers. The assembled crowd continues to heckle, laugh and demean Jesus by hurling spit and foul words. Caiaphas and his council of henchmen are pleased. They do have one gripe: They are mad about the verbiage scrawled on the titulus, the sign above Christ's head. The titulus is used to display the crime He was convicted of as a lesson to all. This one proclaims, "Jesus of Nazareth, King of the Jews." It remains under Pilate's orders.

While most of the crowd is engrossed, some are completely shaken by what they see - an unrecognizable man with a face grossly disfigured and covered in caked blood. Lacerations cover His body. Both eyes are swollen closed. He has massive swelling around His cheek bones and a bent nose. His lips are several sizes bigger than normal. Four large spikes ooze blood from His hands and feet. A dirty, bloody crown is imbedded on His head. A discolored naked body sliced to pieces. Only a few people drop to their knees and weep. The mocking has ceased. It is dead silent.

The heat of the day is cooling over His naked, blood-soaked body. Hypothermia is setting in. Jesus' breathing is shallow and labored. The rise and fall of His chest are scarcely visible. His body is still cramping and spasming uncontrollably. In His shattered state, He is unable to push Himself upward which could lessen His torment. His tongue is swollen and His throat is dry and raw. He is choking and close to seizure and coma. He can no longer bare the relentless pain. Jesus has been suspended for three hours. His organs have failed.

As Christ's life slips away, He looks up to Heaven and proclaims, "It is finished." With a final struggle to speak, he cries out, "Father into your hands I commit my spirit."

Jesus Christ, The Son of God, is dead at the age of 34. It is 3 p.m., April 8, 30 AD.

The Centurion pierces Christ's chest with a spear to confirm His death.

✝

"It Is Finished"

John 19:30 (KJV)

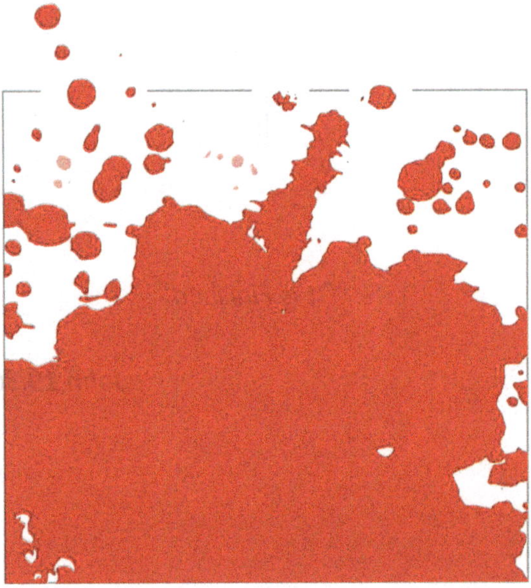

Chapter XIV

CAUSE OF DEATH

Christ was a victim of three long hideous batterings that the average person could not withstand. Imagine being battered by scores of powerful, angry men for 45 minutes. The only rule is to not kill you. Or being flogged by trained assassins, with no limit to the number of lashes other than they can't kill you. Just leave some life for the final executioner. Then, when you are virtually dead on your feet, you are nailed to a cross for three hours.

The number of wounds Jesus sustained was staggering. For the scourging, some estimates exceed two hundred wounds to His torso. The intense beatings of slaps and punches to His head and face were made by literally hundreds of people during His interrogation, fake trial, at the Praetorium and trek to Golgotha. All would have killed any man before being crucified.

The causation of Christ's traumatic death on the cross was an accumulative effect from the damage of all of the tortures He endured. It is likely He lost close to 40 percent of His blood volume during His suffering, including fluid lost from profuse sweating, vomiting and other body secretions. Multiple injuries yielded wide-spread tissue damage that thinned His blood, resulting in more blood loss. Jesus would have experienced coagulopathy, the failure of blood to clot.

Christ's awkward body position on the cross generated crushing pressure and difficulty in breathing - a condition known as positional asphyxiation. Heart arrhythmia will follow these afflictions. In addition to the traumatic shock from all of the blunt force trauma, coupled with severe dehydration, Jesus was pushed into the final stage of Hypovolemic Shock. His organs began to shut down; heart failure and respiratory arrest ensued.

The summary of Jesus' injuries were:

1. Craniofacial injuries: eyes, nose, ears, head
2. Craniomaxillofacial injuries: mouth, teeth, lips, jaw
3. Massive bleeding: internal and external
4. Extreme dehydration: zero fluid intake, profuse sweating, massive bleeding, depletion of electrolytes
5. Head trauma: traumatic brain injury, concussion, second impact injury
6. Organ failure: heart, lungs, kidneys, spleen
7. Positional asphyxiation: lungs and heart
8. Respiratory acidosis: collapsed lung, rib damage
9. Coagulopathy: blood can't clot
10. Hypovolemic shock: enormous fluid loss
11. Traumatic shock: blunt force trauma
12. Surrender of life: Jesus was at the end of His mission and was ready to sacrifice Himself. He had accomplished what God had tasked Him to do.

✝

But He was pierced for our transgressions, He was crushed for our iniquities; the punishment that brought us peace was on Him, and by His wounds we are healed.

Isaiah 53:5 (NIV)

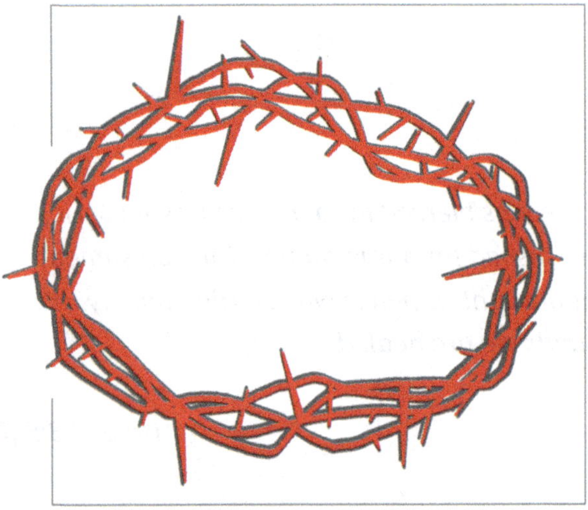

Chapter XV

BODY RECOVERED

Joseph of Arimathea, a disciple and Graves Registration Unit leader, requested from Pilate to retrieve Christ's body. This was granted.

The unit also consisted of Nicodemus, Mary Magdalene and another follower named Mary. As per the units' procedures, Jesus' body was washed and carefully wrapped in linen. He was placed in a tomb and the location was recorded. As per tradition, a huge stone was rolled in front of the tomb. This concluded the burial detail.

Jesus' personal effects should have been recovered and given to His mother. This did not happen because the four-soldier crucifixion detail casted lots, while Christ was still on the cross, and divided up the meager belongings.

✝

For God made Christ, who never sinned, to be the offering for our sin, so that we could be made right with God through Christ.

2 Corinthians 5:21 (NLT)

Chapter XVI

FINAL DEBRIEF

(Christ's Sacrifice)

Jesus died for our sins, even though He had no sins of His own. It was a trade: His one sin-free life and horrific death for our many sin-filled lives and our salvation.

Christ shed His blood and endured immense pain for our transgressions; He was our replacement on the cross.

The horrendous final hours of being scourged, beaten and crucified that slowly destroyed His mortal being was the worst penalty ever. Only an ultimate atonement that left Him with nothing else to give would qualify as a large enough sacrifice to make it an equal trade for the enormity of all of mankind's sins. The seemingly senseless brutality of Christ's crucifixion now makes perfect sense when contrasted with the size and gravity of the trade.

We live in a violent society where war, murders, crimes, hate and assaults are everyday headlines. Sadly, we have been desensitized from the daily onslaught of our own violent ways. As a result, we gloss over, delete and forget the gruesome details in how Jesus died in exchange for our salvation. We must never forget that Christianity is about the crucifixion and all its barbarity.

By His wounds, we were saved.

"He Himself bore our sins" in His body on the cross, so that we might die to sins and live for righteousness; by His wounds you have been healed."

1 Peter 2:24 (NIV)

Christ's Purple Heart is distinguished by the cross on which He died. Above the heart is a shield, which symbolizes Jesus as the protector. It is red for the sizeable amount of blood He shed and is flanked with thorns representing the crown and various devices used to inflict His injuries. The gold "V" that lies atop the shield stands for the incredible valor Christ exhibited while being tortured as a prisoner of war. The red ribbon trimmed in purple is for additional blood He shed, more than any man could ever endure.

Chapter XVII

CHRIST'S PURPLE HEART

(Posthumously)

to

Jesus Christ

For wounds received in action against hostile enemy forces resulting in His death, while engaged in a world-changing mission April 30 AD, Jerusalem, Judea.

Orchestrating all activities against a large combat detachment, He single handedly conducted and commanded all operations successfully despite having inferior numbers. Seized during an ambush, He was held captive and interrogated repeatedly. He was severely tortured for more than eighteen hours. He was scourged and beaten. His face and body were battered beyond recognition. He refused to cooperate or provide any information that might compromise his mission or jeopardize the safety of His men. Though death was imminent, He continued to remain silent and was executed by crucifixion, sacrificing His life for all.

The Valor "V" Device is authorized.

For Christ also hath once suffered for sins, the just for the unjust, that He might bring us to God; being put to death in the flesh, but quickened by the Spirit.

1 Peter 3:18 (KJV)

REFERENCES

Holy Bible translations: English Standard Version (ESV), King James Version (KJV), New American Standard Bible (NASB), New International Version (NIV), New King James Version [NKJV], New Living Translation (NLT)

Bergeron, Joseph. *The Crucifixion of Jesus.* Joseph Bergeron, 2018.

Bishop, Jim. *The Day Christ Died.* Harper Collins, 1977.

Chapman, David. *Ancient Jewish and Christian Perceptions of Crucifixion.* Baker Publishing Co., 2008.

Chilton, Bruce. *The Herods: Murder, Politics and the Art of Succession.* Fortress Press, 2021.

Constantinou, Eugenia. *The Crucifixion of the King of Glory.* Ancient Faith Publishing, 2021.

Coulter, Fred. *The Day Jesus the Christ Died.* York Publishing, 2011.

De La Bedoyere, Guy. *Gladius: The World of the Roman Soldier.* University of Chicago Press, 2020.

Dixon, Tris. *Damage the Untold: Story of Brain Trauma in Boxing.* Hamilcar Publications, 2021.

Duriez, Colin. *The Year that Changed the World: AD 33.* Intervarsity Press, 2006.

Emmerich, Anne. *The Dolorous Passion of Our Lord Jesus Christ.* Tan Books, 1983.

Fuhrman, Christopher. *Policing the Roman Empire.* Oxford University Press, 2012.

Gorman, Ralph. *The Last Hours of Jesus.* St. Pius X Press, 2022.

Graves, David. *The Cities of the New Testament World.* Independently Published, 2023.

Hengel, Martin. *Crucifixion.* Fortress Press, 1977.

Hesemann, Michael. *Jesus of Nazareth.* Ignatius Press, 2021

Jensen, Robin. *The Cross.* Harvard University Press, 2017

Jeremias, Joachim. *Jerusalem in the Time of Jesus.* Fortress Press, 1969.

Kennedy, Titus. *Excavating the Evidence for Jesus.* Harvest House Publishers, 2022.

Kubala, Mark. *The Execution of Jesus the Christ.* West Bow Press, 2017.

Macarthur, John. *The Murder of Jesus.* Nelson Books, 2004.

McDowell, Sean. *The Fate of the Apostles.* Taylor and Fracis Group, 2018.

McGovern, Thomas. *What Christ Suffered.* Our Sunday Visitor Publishing, 2021.

Miller, Stephen. *Eyewitness to Crucifixion.* Our Daily Bread Publishing, 2020.

Ratzinger, Joseph. *Jesus of Nazareth: Volume II.* Ignatius Press, 2011.

Rutledge, Fleming. *The Crucifixion.* WM.B. Erdmans Publishing, 2015.

Rutledge, Steven. *The Death of Christ.* Pen & Sword History, 2022

Schnabel, Eckhard. *Jesus in Jerusalem: The Last Days.* WM. B. Eedermans Publishing Co., 2018.

Schwind, David. *Sacrifice Remembered: Posthumous Awards of the Purple Heart.* Medals Press, 2019.

Siebeck, Mohr. *Crucifixion in the Mediterranean World.* Mohr Siebeck, 2019.

Sri, Edward. *No Greater Love: A Biblical Walk Through Christ's Passion.* Ascension Publishing, 2019.

Stott, John. *The Cross of Christ.* Intervarsity Press, 2021.

Strobel, Lee. *The Case for Easter.* Zondervan, 2003.

Unterharnscheidt, Friedrich and Julia Taylor. *Boxing Medical Aspects.* Academic Press, 2003.

Wall, J. Charles. *Relics from the Crucifixion.* Sophia Institute Press, 2015.

West, William. *Riddles of the Shroud.* William West, 2022.

Zeichmann, Christopher. *The Roman Army and the New Testament.* Lexington Books Fortress Academic, 2018.

Zugibe, Frederick. *The Crucifixion of Jesus: A forensic Inquiry.* M. Evans, 2005.

References

ABOUT THE AUTHOR

Chuck Habermehl is a decorated U.S. Marine Vietnam Veteran who served his country from 1971 to 1975. During his 38 years in law enforcement, he achieved the rank of Deputy Chief in an active urban jurisdiction of Miami-Dade County.

In 1994, he founded the training company Direct Action Defensive Systems, followed by the CQB-Close Quarters Battle, Inc., and CQB, LLC, headquartered in Florida. Habermehl has trained thousands of police and military personnel inside and outside the U.S., including many top-tier elements.

Habermehl has decades of experience in various hand-to-hand fighting systems; He managed a Marine Corps gym facility; spent 10 years as a police academy senior defensive tactics instructor at the Southeast Florida Institute of Criminal Justice in Dade County, Florida; and 16 years as Director of Training at Volusia County Sheriff's Office before retiring in 2016.

Habermehl developed CQB training enhancements adopted by many organizations and has written and produced several tactical training manuals, books and videos reserved for LEO and military units. His 2011 book *Depictions: Poems About Warriors and War*, Habermehl's first book made available to the public at the time, was awarded a gold medal by the Military Writers Society of America.

Habermehl is a graduate of the University of Miami.

Made in the USA
Monee, IL
03 November 2026

67383331R00075